JUMBO COLORING BOOK

A DESERT ECOSYSTEM IS A DRY BUT RICH HABITAT WHERE PLANTS AND ANIMALS HAVE UNIQUELY ADAPTED TO SURVIVE HARSH CONDITIONS, SUPPORTING A SURPRISING VARIETY OF LIFE.

LITTLE ARTIST STUDIO, A REGISTERED DBA OF SOPHIA N. JOHNSON, LLC. ALL RIGHTS RESERVED. REGISTERED IN NEW YORK STATE.

Biographical Note
Desert Ecosystem An Educational Coloring Book is a new work,
first published by Little Artist Studio in 2025.

International Standard Book Number
ISBN 979-8-9992504-4-5

www.littleartiststudio.org

DISCOVER THE STUNNING BEAUTY OF THE WORLD'S DESERT ECOSYSTEMS WITH OVER 85 INTRICATELY DETAILED COLORING PAGES. THIS EDUCATIONAL COLORING BOOK HIGHLIGHTS ICONIC DESERTS FROM AROUND THE GLOBE, DESIGNED TO INSPIRE CURIOSITY AND DEEPEN UNDERSTANDING OF THESE UNIQUE AND VITAL ENVIRONMENTS. AS PART OF LITTLE ARTIST STUDIO'S ACCLAIMED EDUCATIONAL SERIES, EACH FULL-PAGE ILLUSTRATION TELLS A POWERFUL STORY THAT FUELS CREATIVITY AND LEARNING. FEATURING SINGLE-SIDED PAGES, ARTISTS OF ALL AGES CAN EXPERIMENT WITH ANY MEDIUM AND EASILY SHOWCASE THEIR FINISHED MASTERPIECES. IDEAL FOR NATURE ENTHUSIASTS, EDUCATORS, AND CREATIVE EXPLORERS ALIKE.

DESERT

A DESERT IS A TYPE OF ECOSYSTEM THAT GETS VERY LITTLE PRECIPITATION- USUALLY LESS THAN 10 INCHES OF RAIN OR SNOW EACH YEAR.

DESERT BIOME

THE DESERT BIOME COVERS ABOUT ONE-FIFTH
OF EARTH'S LAND. ITS GROUND CAN BE
SANDY, ROCKY, OR COVERED WITH SMALL
STONES, DEPENDING ON THE TYPE OF DESERT.

DESERT BIOME

DESERTS ARE VERY DRY PLACES WHERE
ONLY CERTAIN PLANTS AND ANIMALS CAN
SURVIVE.

ARMADILLO LIZARD

DESERT BIOME

DESERT PLANTS HAVE SPECIAL WAYS TO SAVE WATER. FOR EXAMPLE, CACTI STORE WATER IN THEIR THICK STEMS AND HAVE SHARP SPINES TO KEEP ANIMALS FROM TAKING IT.

DESERT BIOME

ANIMALS SUCH AS THE BLACK-TAILED JACKRABBIT (LEPUS CALIFORNICUS) ARE ALSO ADAPTED TO LIFE IN THE DESERT.

DESERT BIOME

THE BLACK-TAILED JACKRABBIT HAS BIG
EARS THAT HELP IT STAY COOL BY LETTING
HEAT ESCAPE INTO THE AIR.

RED-TAILED HAWK

FOUR TYPES OF DESERTS

THE FOUR MAIN TYPES OF DESERT INCLUDE:
SUB-TROPICAL (HOT AND DRY DESERTS),
SEMI-ARID DESERTS, COASTAL DESERTS, AND
POLAR (COLD DESERTS).

SUBTROPICAL DESERTS

SUBTROPICAL OR HOT AND DRY DESERTS,
LIKE THE SAHARA, ARE VERY HOT DURING
THE DAY AND GET ALMOST NO RAIN.

SUBTROPICAL DESERTS

SUBTROPICAL DESERTS ARE PRIMARILY FOUND BETWEEN 15° AND 30° NORTH AND SOUTH OF THE EQUATOR, WHICH IS ROUGHLY CENTERED ON THE TROPICS OF CANCER AND CAPRICORN, BUT NOT ALONG THE EQUATOR ITSELF.

SUBTROPICAL DESERTS

THE AIR IS DRY, AND THE LAND IS SANDY OR ROCKY.

SUBTROPICAL DESERTS

ADDAX ANTELOPES ARE NATIVE TO THE
SAHARAN DESERT.

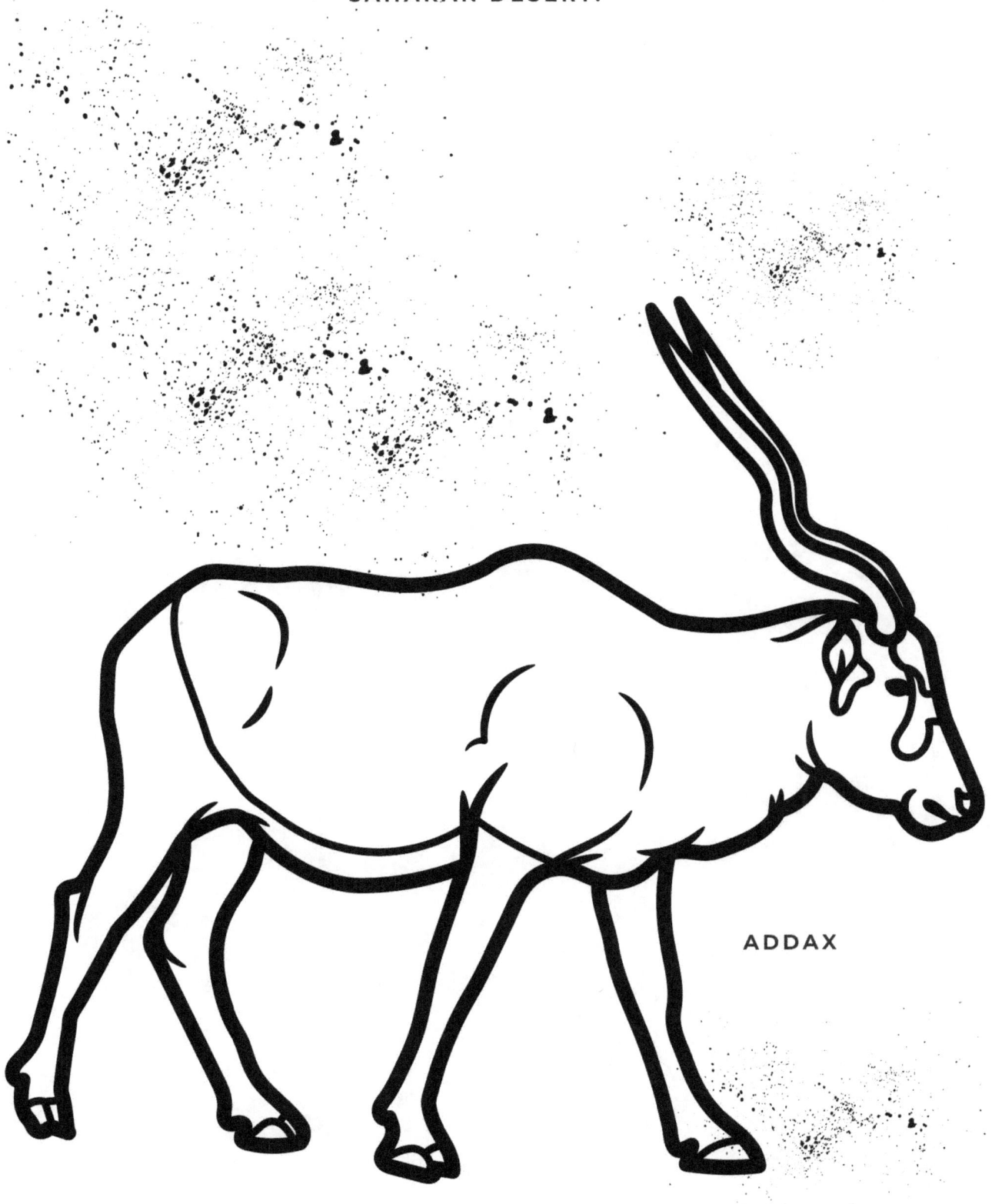

ADDAX

SEMI-ARID DESERTS

SEMI-ARID DESERTS AREN'T AS HOT AND GET A LITTLE MORE RAIN.

SEMI-ARID DESERTS

SEMI-ARID DESERTS ARE LOCATED IN ASIA, EUROPE AND NORTH AMERICA.

SEMI-ARID DESERTS

SOME GRASSES AND BUSHES CAN GROW IN
SEMI-ARID DESERTS.

SAGE

COASTAL DESERTS

COASTAL DESERTS LIKE THE ATACAMA
DESERT IN CHILE, ARE NEAR THE OCEAN BUT
STILL VERY DRY.

COASTAL DESERTS

COASTAL DESERTS FORM ALONGSIDE THE TROPICAL WESTERN EDGES OF PARTS OF SOUTH AMERICA AND AFRICA - AND DUE TO COLD OCEAN CURRENTS.

COASTAL DESERTS

NAMIBIA'S COASTAL DESERT IS WHERE TALL SAND DUNES MEET THE SEA. IT'S A SPECIAL PLACE IN SOUTHWEST AFRICA AND A UNESCO WORLD HERITAGE SITE!

COASTAL DESERTS

COASTAL DESERTS CAN BE COOL AND FOGGY,
WITH LITTLE RAIN.

POLAR DESERTS

POLAR OR COLD DESERTS, LIKE ANTARCTICA,
ARE FREEZING AND SNOWY BUT GET VERY
LITTLE RAIN OR SNOW EACH YEAR.

POLAR DESERTS

POLAR DESERTS ARE FOUND IN THE ARCTIC AND ANTARCTIC, WHERE WINTER TEMPERATURES AVERAGE BELOW FREEZING.

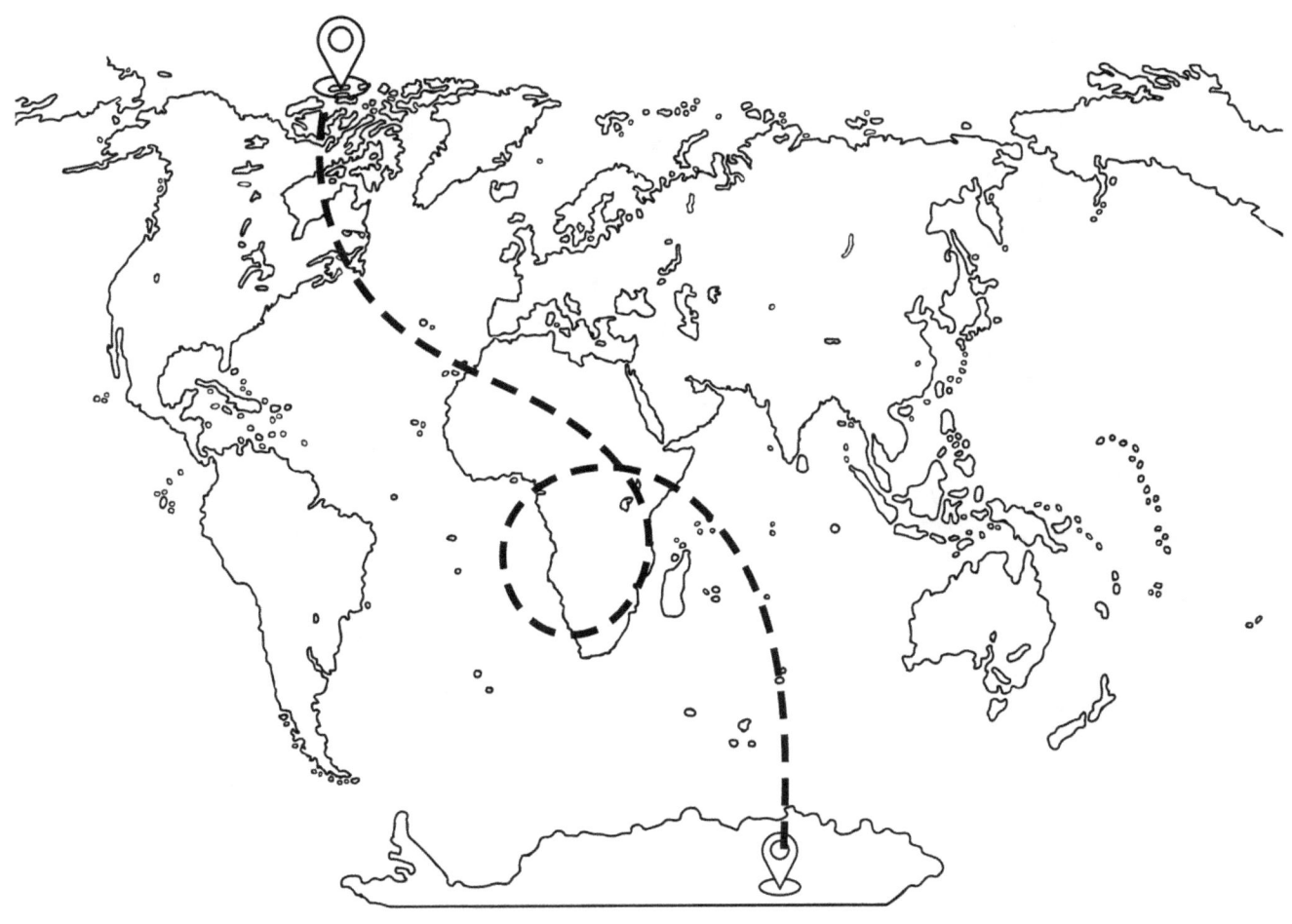

POLAR DESERTS

POLAR DESERTS ARE DRY JUST LIKE HOT DESERTS!

PLANTS AND ANIMALS

EVEN THOUGH DESERTS ARE DRY, THEY'RE
FULL OF SMART PLANTS AND ANIMALS
THAT KNOW HOW TO SURVIVE!

PLANTS AND ANIMALS

SOME COMMON DESERT PLANTS
INCLUDE CACTI, SUCCULENTS, AND
SHRUBS LIKE CREOSOTE.

PLANTS AND ANIMALS

FENNEC FOXES HAVE BIG EARS THAT HELP THEM STAY COOL. THE HEAT FROM THEIR BODIES GOES OUT THROUGH THEIR EARS LIKE A FAN!

PLANTS AND ANIMALS

SOME DESERT PLANTS HAVE TINY
LEAVES OR SHARP SPINES TO STOP
WATER FROM DRYING UP. THIS HELPS
THEM STAY WET AND HEALTHY IN THE
HOT SUN!

SUCCULENT FLOWER

PLANTS AND ANIMALS

PLANTS AND ANIMALS

THIS DROMEDARY (ONE-HUMPED) CAMEL STORES
FAT IN ITS HUMP AND USES IT FOR ENERGY AND
WATER WHEN THEY NEED IT.

PLANTS AND ANIMALS

MEERKATS ARE DESERT ANIMALS
THAT LIVE IN GROUPS AND WATCH
OUT FOR DANGER TOGETHER.

PLANTS AND ANIMALS

AGAVE PLANTS HAVE THICK LEAVES THAT STORE
WATER. THEY BLOOM ONE BIG FLOWER IN THEIR LIFE,
THEN DIE. THEY LIVE WELL IN HOT, DRY DESERTS!

PLANTS AND ANIMALS

RATTLESNAKES USE JUST A LITTLE
VENOM WHEN THEY NEED IT-SO THEY
DON'T RUN OUT IN THE DESERT!

PLANTS AND ANIMALS

THE DESERT KANGAROO RAT IS A
SMALL ANIMAL THAT LIVES IN DRY
PLACES AND GETS ALMOST ALL ITS
WATER FROM THE SEEDS IT EATS.

PLANTS AND ANIMALS

THIS TORTOISE CAN SURVIVE IN EXTREME HEAT BY DIGGING
BURROWS TO STAY COOL AND CONSERVING WATER IN ITS BODY.

PLANTS AND ANIMALS

DESERT SCORPIONS ARE SMALL, VENOMOUS CREATURES THAT LIVE IN DRY, HOT DESERTS AND COME OUT AT NIGHT TO HUNT.

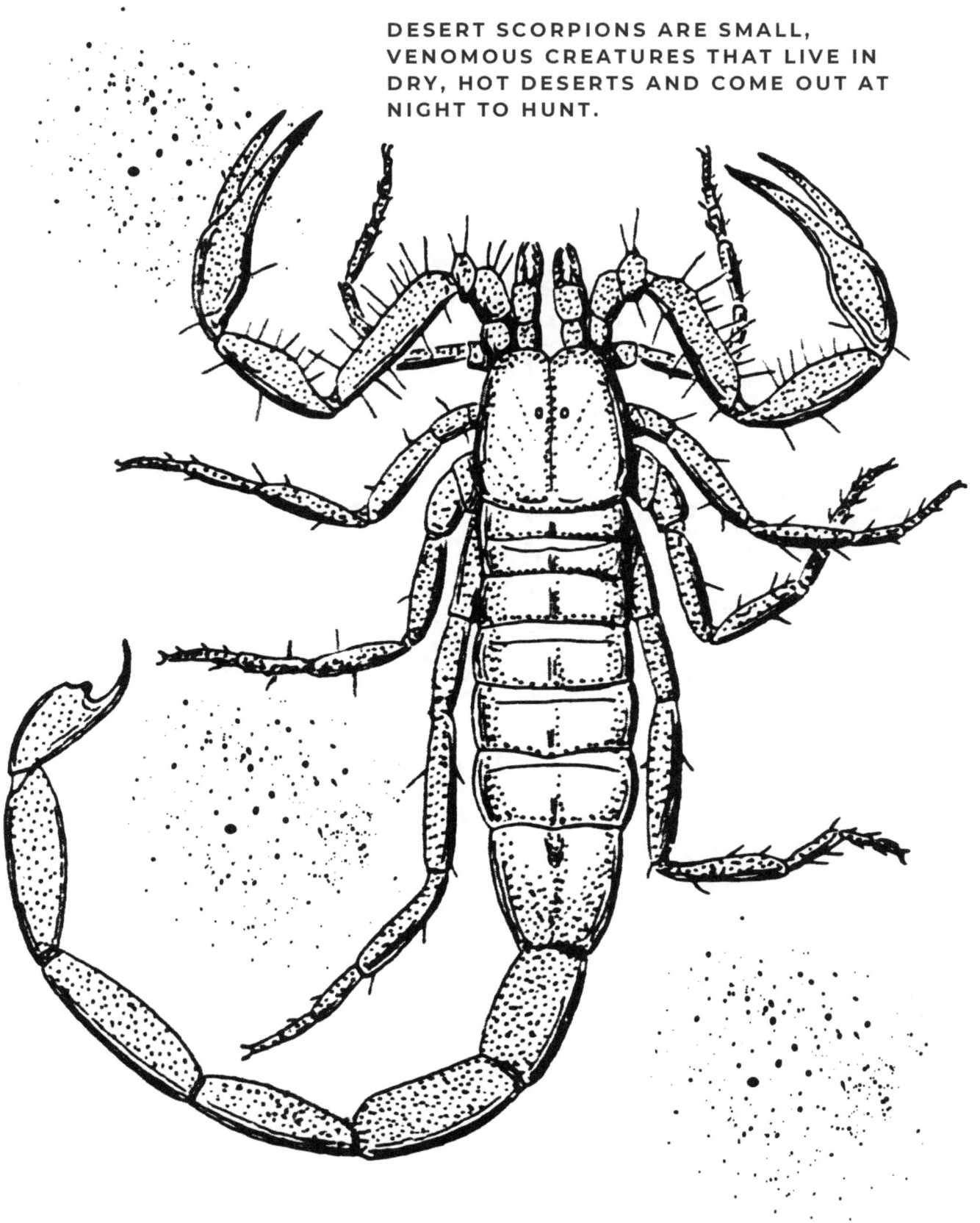

PLANTS AND ANIMALS

CAMELS ARE KNOWN AS THE "SHIP OF THE DESERT," THIS BACTRIAN (TWO-HUMPED) CAMEL CAN TRAVEL LONG DISTANCES AND CARRY HEAVY LOADS.

PLANTS AND ANIMALS

THE GILA MONSTER IS A
DESERT LIZARD THAT STORES
FAT AND WATER IN ITS TAIL.

PLANTS AND ANIMALS

THE SIDEWINDER RATTLESNAKE, GET MOISTURE
FROM THE FOOD THEY EAT. THEIR SCALES HELP
PROTECT THEM FROM THE HEAT OF THE SAND.

PLANTS AND ANIMALS

COYOTES ARE NOCTURNAL, WHICH HELPS THEM AVOID THE HEAT OF THE DAY.

PLANTS AND ANIMALS

A CHUCKWALLAS IS A TYPE OF
LIZARD THAT LIVES IN THE
DESERTS OF NORTH AMERICA.

PLANTS AND ANIMALS

HORNED LIZARDS
CAN PUFF ITSELF UP
TO LOOK BIGGER AND
DEFEND ITSELF.

SAHARA DESERT

THE SAHARA DESERT USED TO BE GREEN
AND FULL OF LAKES AND ANIMALS, BUT
NOW IT'S THE BIGGEST HOT DESERT IN
THE WORLD!

SAHARA DESERT

A LONG TIME AGO, ANIMALS LIKE ELEPHANTS, GAZELLES, RHINOS, AND GIRAFFES LIVED IN THE SAHARA DESERT, AND DRANK FROM THE WATER.

SAHARA DESERT

IF YOU DIG DEEP, YOU'LL FIND HARD BEDROCK
AND SOMETIMES DRY, CRACKED CLAY HIDING
UNDERNEATH ALL THAT DESERT SAND.

SAHARAN OASIS

A SAHARAN OASIS IS A GREEN, WET PLACE IN THE MIDDLE OF THE HOT, DRY SAHARA DESERT!

ATACAMA DESERT

THE ATACAMA DESERT IN SOUTH AMERICA IS ONE OF THE DRIEST PLACES ON EARTH, WITH SOME SPOTS GETTING NO RAIN FOR HUNDREDS OF YEARS!

ARGENTINIAN
PUMA

ATACAMA DESERT

THE ATACAMA DESERT IS LOCATED IN CHILE. IT'S SO DRY AND ROCKY THAT IT LOOKS A LOT LIKE MARS! SCIENTISTS GO THERE TO STUDY WHAT LIFE MIGHT BE LIKE ON OTHER PLANETS AND TO TEST SPACE TOOLS THAT MIGHT BE USED ON MARS.

ATACAMA DESERT

LLAMAS DIDN'T ORIGINALLY LIVE IN CHILE, BUT PEOPLE BROUGHT THEM THERE. NOW, YOU CAN FIND THEM LIVING IN THE HIGH, DRY PLAINS IN THE NORTH OF THE COUNTRY.

ATACAMA DESERT

THE OASIS HUMMINGBIRD IS A SMALL, COLORFUL BIRD THAT LIVES IN DRY PLACES LIKE THE ATACAMA DESERT IN CHILE AND PERU.

ATACAMA DESERT

EVEN THOUGH THE DESERT IS VERY DRY, THE HUMMINGBIRD FINDS WATER AND FLOWERS IN SPECIAL SPOTS CALLED OASES, WHERE IT CAN DRINK NECTAR AND STAY SAFE.

BLUE AGAVE

ATACAMA DESERT

THE OASIS HUMMINGBIRD IS THE SOLE SPECIES IN THE GENUS RHODOPIS, WHICH IS WHY THE GENUS IS CONSIDERED MONOTYPIC.

ATACAMA DESERT

THE OASIS HUMMINGBIRD IS PART OF A SPECIAL HUMMINGBIRD GROUP CALLED THE "BEE HUMMINGBIRDS" BECAUSE THEY ARE SMALL LIKE BEES!

RAIN SHADOW DESERT

A RAIN SHADOW DESERT FORMS ON THE DRY SIDE OF A MOUNTAIN. WHEN MOIST AIR GOES UP ONE SIDE OF THE MOUNTAIN, IT RAINS THERE. BY THE TIME THE AIR GETS TO THE OTHER SIDE, IT'S DRY-MAKING A DESERT!

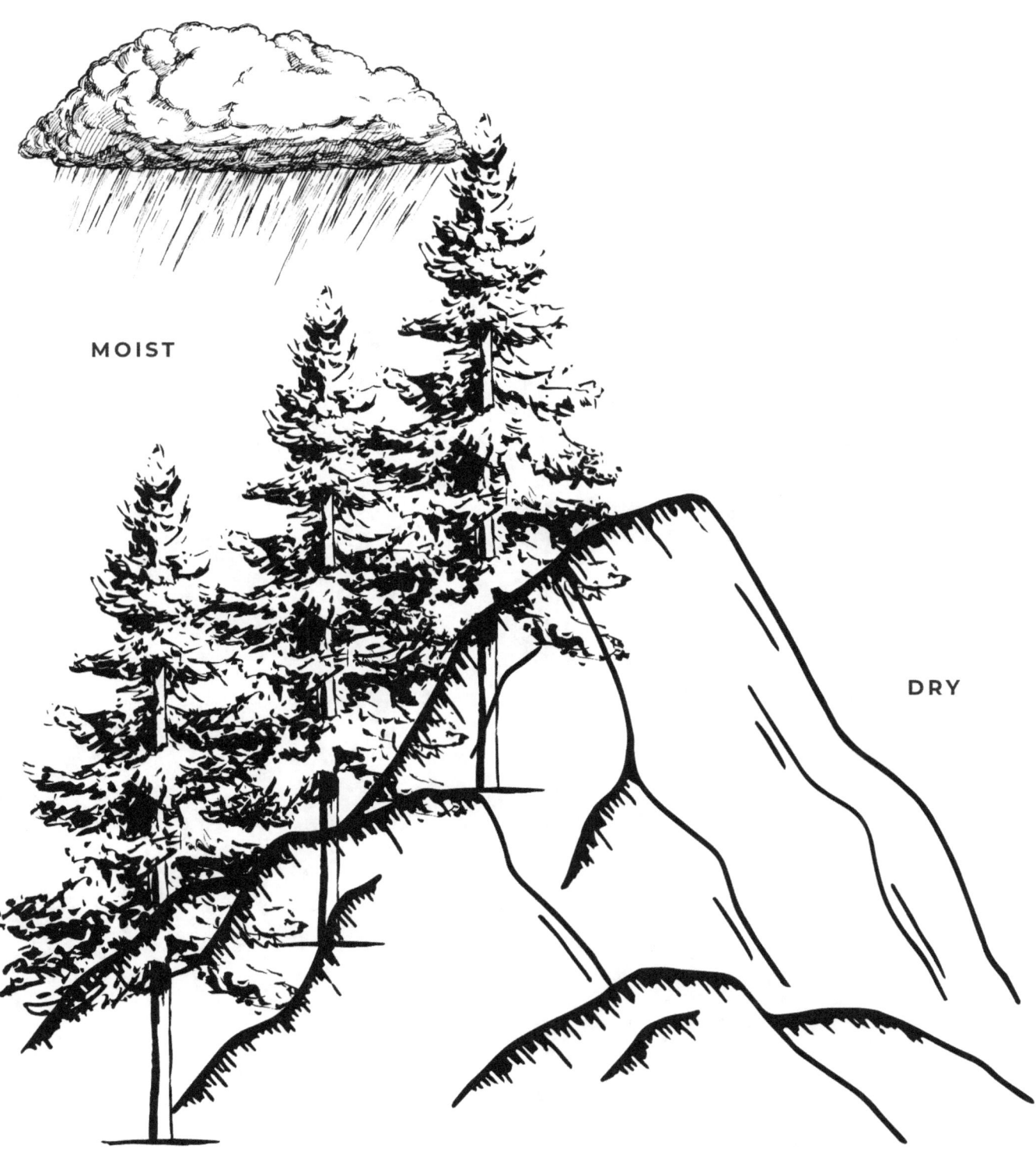

MOIST

DRY

DEATH VALLEY

DEATH VALLEY, IN CALIFORNIA AND NEVADA,
IS A SPECIAL KIND OF DESERT CALLED A RAIN
SHADOW DESERT, WHERE MOUNTAINS BLOCK
THE RAIN FROM GETTING IN.

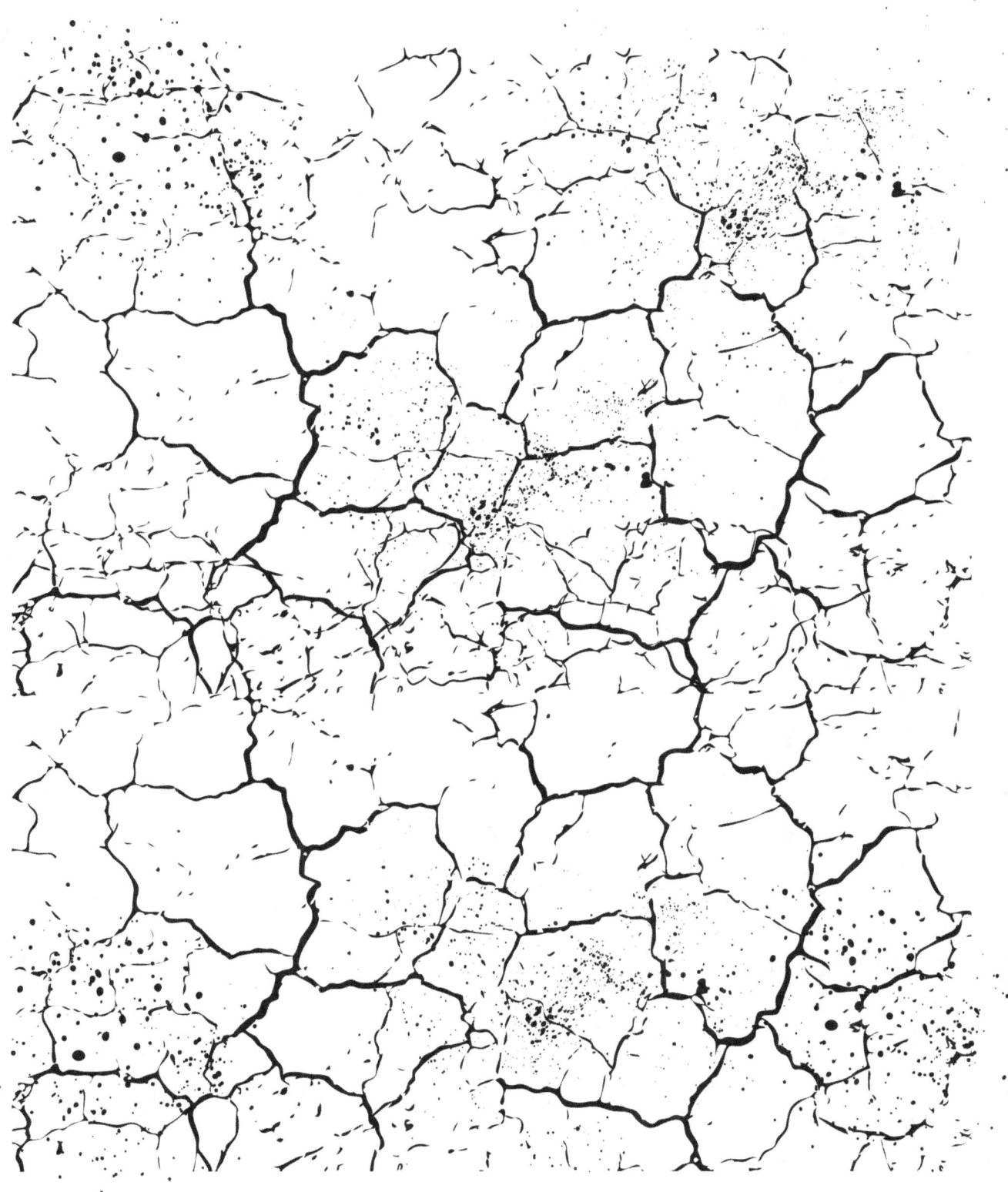

GOBI DESERT

THE GOBI DESERT IS LOCATED IN ASIA, SPANNING NORTHERN CHINA AND SOUTHERN MONGOLIA. THE GOBI EXPERIENCES EXTREMELY COLD TEMPERATURES, ESPECIALLY DURING WINTER, AND SNOW CAN OCCUR.

THE GHAGGAR RIVER

THE GHAGGAR RIVER FLOWS INTO AND
EVENTUALLY DISAPPEARS INTO THE
THAR DESERT IN RAJASTHAN, INDIA.

BLACK ROCK DESERT

THE BLACK ROCK DESERT IN NEVADA IS
WHAT'S LEFT OF A HUGE, ANCIENT
LAKE CALLED LAKE LAHONTAN.

SANDSTORM

DESERT SANDSTORMS CAN COVER
EVERYTHING IN THEIR WAY, LIKE ROCKS,
FIELDS, AND EVEN WHOLE TOWNS!

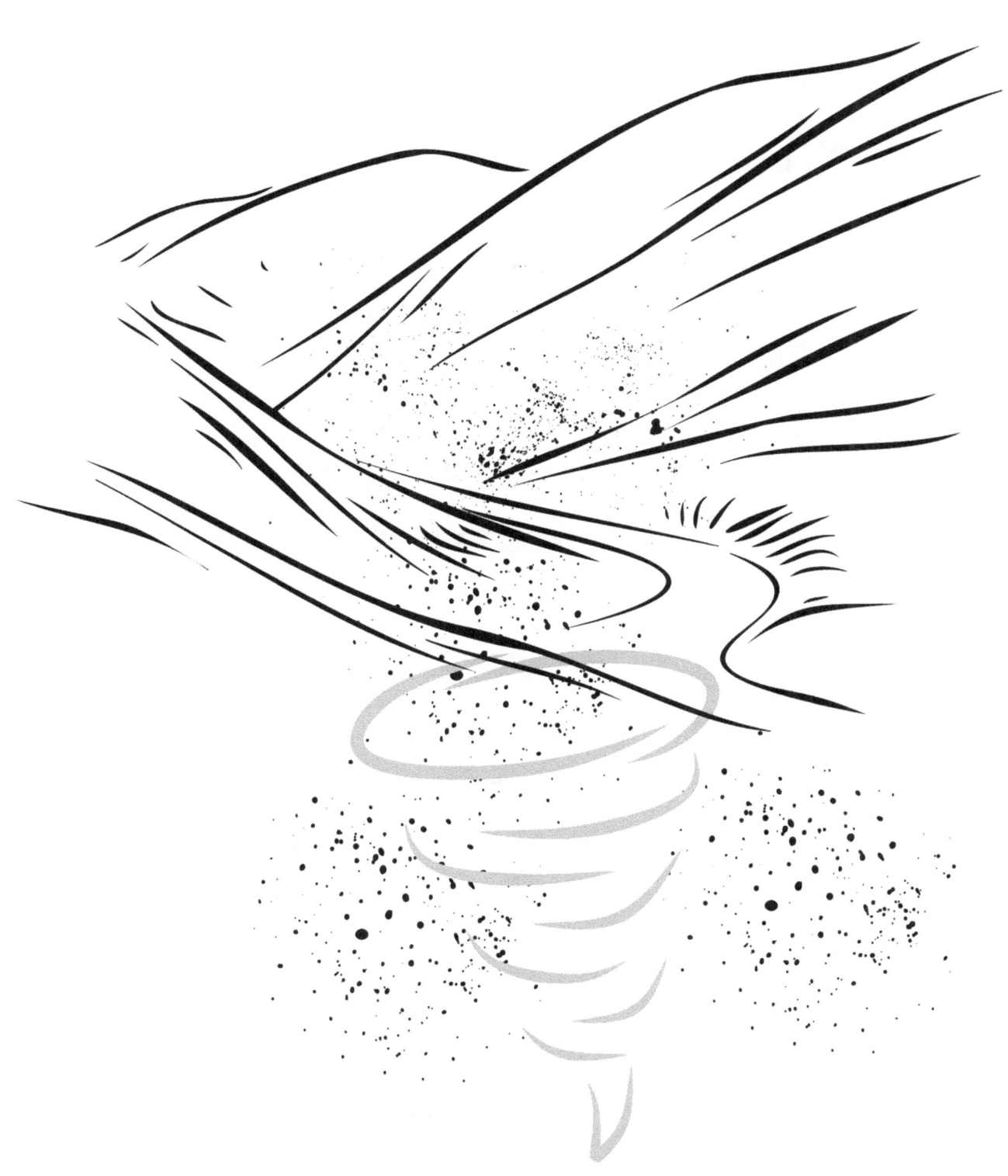

COLORADO PLATEAU DESERT

ARCHES NATIONAL PARK IS IN UTAH. IT HAS LOTS OF
BIG ROCK ARCHES MADE BY WIND AND RAIN. EVEN
THOUGH IT'S A DESERT, IT CAN GET COLD AND EVEN
SNOW!

PATAGONIA

THE DESERTS IN PATAGONIA, THE LARGEST IN
SOUTH AMERICA, ARE GROWING BECAUSE OF
DESERTIFICATION.

PATAGONIA

MORE THAN 30 PERCENT OF THE GRASSLANDS
IN ARGENTINA, CHILE, AND BOLIVIA ARE AT
RISK OF DESERTIFICATION.

TENGGER DESERT

THE TENGGER DESERT IS THE FOURTH-LARGEST DESERT IN CHINA, COVERING ABOUT 36,700 SQUARE KILOMETERS - AND IS FULL OF SAND DUNES.

CHIHUAHUAN DESERT

IN THE CHIHUAHUAN DESERT, IT GETS REALLY HOT
DURING THE DAY AND VERY COLD AT NIGHT. ANIMALS
LIKE THE KANGAROO RAT STAY IN BURROWS TO KEEP
COOL WHEN IT'S HOT AND WARM WHEN IT'S COLD.

LIFE IN THE DESERT

DROMEDARY CAMELS, FROM THE ARABIAN AND SAHARA DESERTS, CAN LOSE UP TO 30% OF THEIR BODY WEIGHT AND STILL BE SAFE.

LIFE IN THE DESERT

SAND LIZARDS, FOUND IN THE DESERTS OF EUROPE AND ASIA, ARE CALLED "DANCING LIZARDS" BECAUSE THEY LIFT ONE LEG AT A TIME TO AVOID THE HOT SAND.

DESERT NOMADS

NOMADIC OR SEMI-NOMADIC PEOPLE LIVE IN
DESERT ECOSYSTEMS BY MOVING FROM
PLACE TO PLACE TO FIND WATER, FOOD,
AND GRAZING LAND FOR THEIR ANIMALS

DESERT NOMADS

NOMADS MOVE OFTEN SO THEIR
SHEEP AND GOATS CAN FIND WATER
AND GRASS TO EAT.

DIGITAL NOMAD

DESERT ECONOMICS

DESERT ECONOMICS IS ALL ABOUT
HOW PEOPLE LIVE, WORK, AND USE
THE LAND IN DRY PLACES!

DESERT ECONOMICS

EVEN THOUGH DESERTS ARE HOT AND
DRY, PEOPLE STILL LIVE THERE AND
FIND SMART WAYS TO WORK!

DESERT ECONOMICS

IN THE DESERT, PEOPLE USE DRIP
IRRIGATION TO GROW FOOD. IT GIVES
LITTLE DROPS OF WATER TO PLANTS SO
THEY DON'T WASTE ANY!

DESERT ECONOMICS

DESERTS GET A LOT OF SUNSHINE, SO PEOPLE USE THE SUN'S LIGHT TO MAKE SOLAR ENERGY. THIS ENERGY HELPS POWER HOMES AND SCHOOLS!

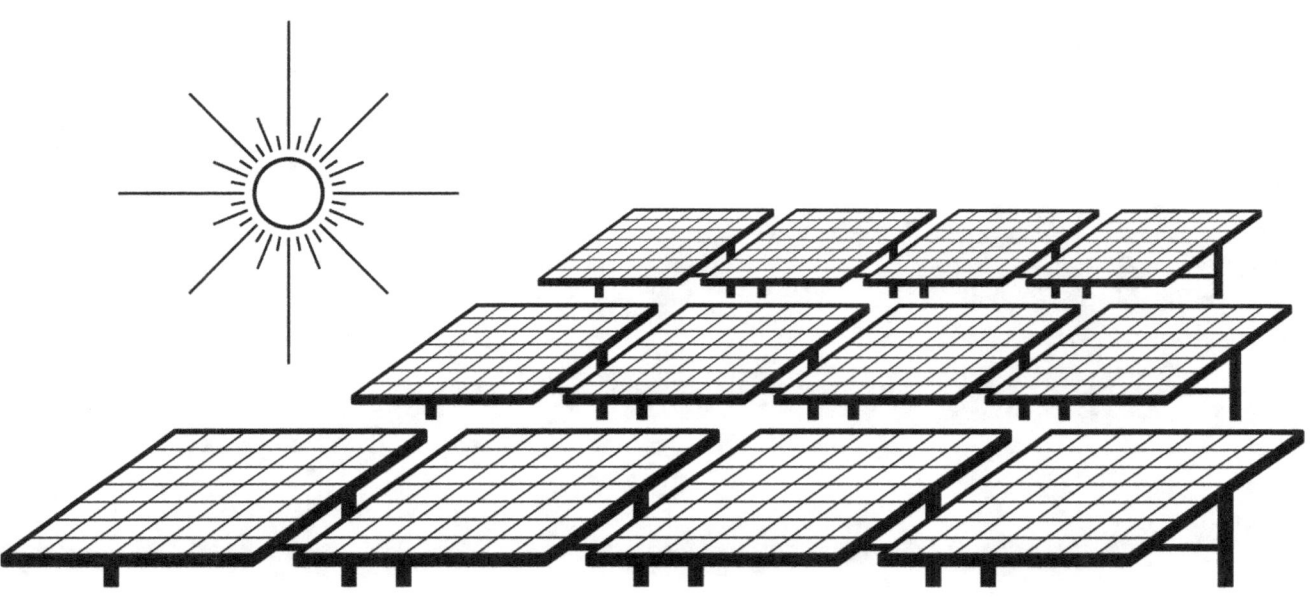

DESSERTIFICATION

DESERTIFICATION OCCURS WHEN LAND THAT USED TO GROW CROPS TURNS INTO DRY, DESERT-LIKE GROUND.

DESSERTIFICATION

WHEN ANIMALS GRAZE TOO MUCH AND TREES ARE CUT DOWN, THE PLANTS THAT KEEP THE SOIL IN PLACE ARE LOST, LEADING TO DEFORESTATION.

DUST BOWL

IN THE 1930S, THE GREAT PLAINS BECAME THE "DUST BOWL" BECAUSE OF DROUGHT AND BAD FARMING, FORCING MILLIONS OF PEOPLE TO LEAVE THEIR FARMS.

DUST DEVILS

A DUST DEVIL IS A SMALL, HARMLESS
WHIRLWIND OF DUST THAT SPINS ACROSS
HOT, DRY PLACES LIKE DESERTS.

CURIOUS FACT

SOME DESERT PLANTS, LIKE THE CREOSOTE
BUSH, CAN LIVE FOR OVER 10,000 YEARS -
MAKING THEM SOME OF THE OLDEST LIVING
ORGANISMS ON EARTH!

CREOSOTE BUSH

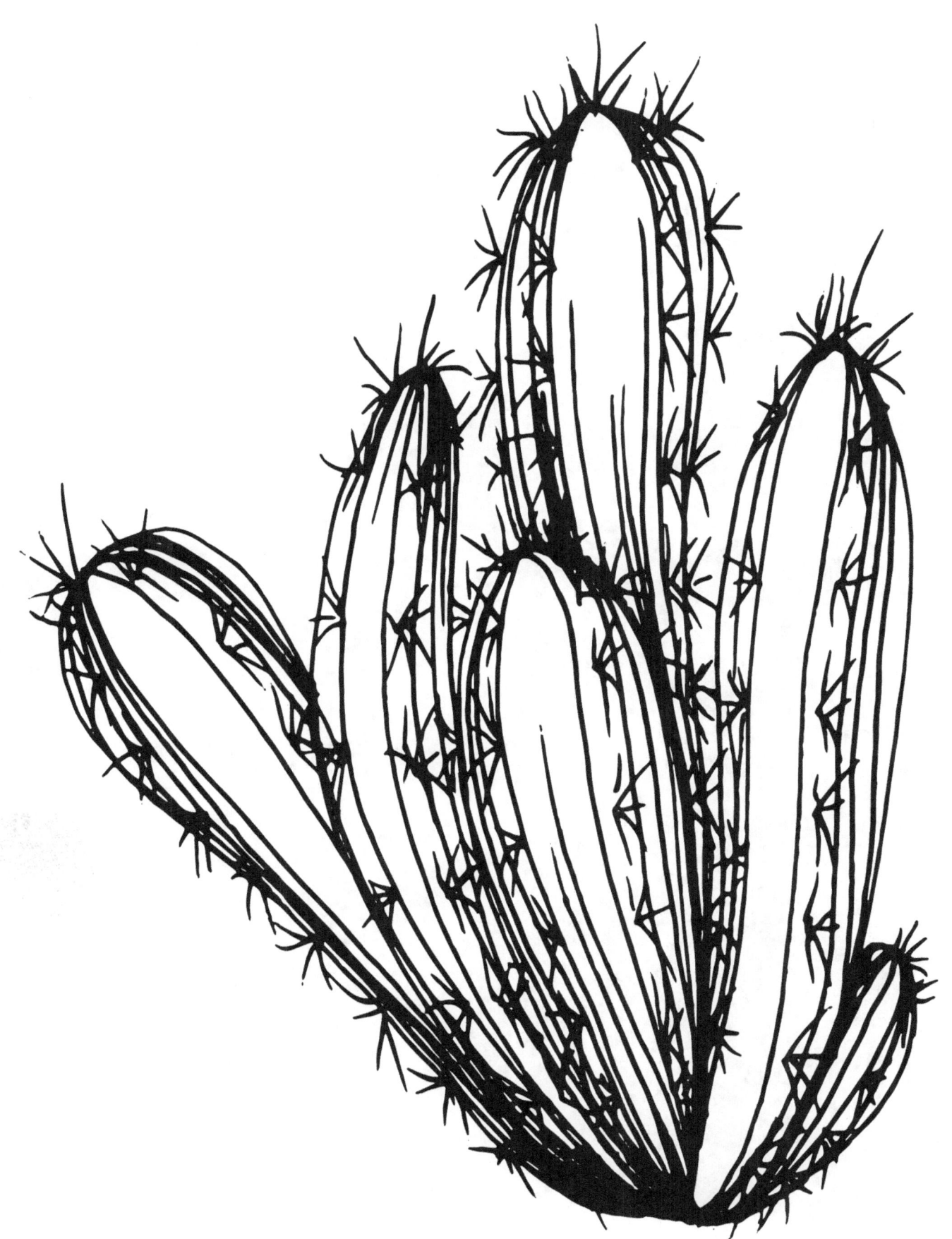

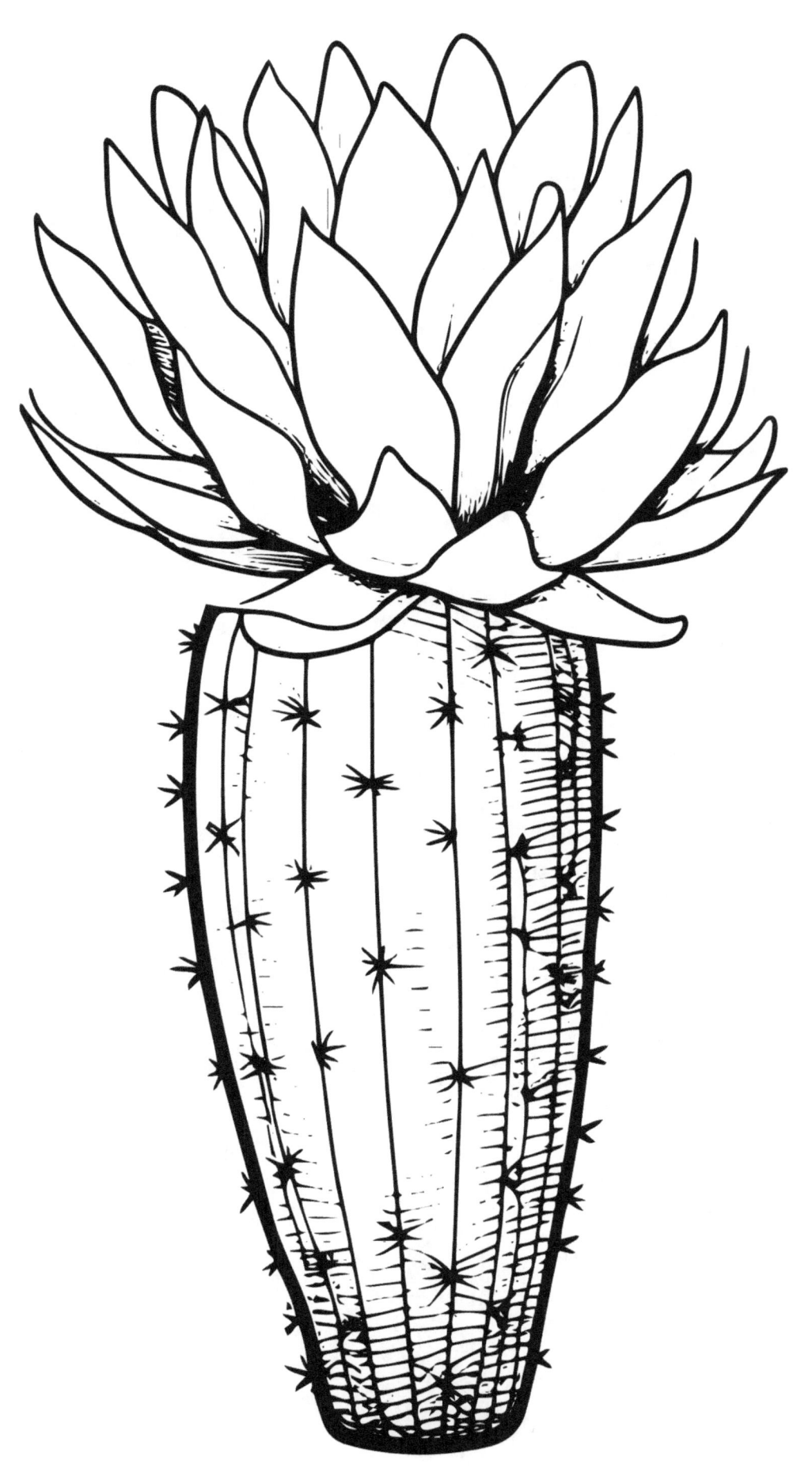

DRAW YOUR OWN DESERT PLANT

"THE BEAUTY OF THE DESERT LIES NOT
IN WHAT YOU SEE, BUT IN WHAT
PATIENTLY WAITS TO BE DISCOVERED."

-AUTHOR UNKNOWN

NOTES